333 Grand Prix Facts Every Motorsport Fan Must Know: The Ultimate Fact Book

Reece Frost

Contents

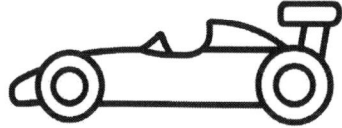

Chapter One

Engines That Roar: The Science of Speed

Did you know that a Formula 1 engine can rev up to 15,000 times per minute? That means the pistons inside the engine are moving so fast they could travel up and down more than 250 times in just one second! That's faster than you can blink ten times in a row. The sound isn't just noise—it's the heartbeat of pure racing power.

Here's something wild: F1 cars don't just burn regular gas like your family's car. They use a special high-performance fuel that's a lot like jet fuel. It's carefully tuned to burn faster, hotter, and more efficiently, and every drop counts. Teams are only allowed 110 kilograms (about 145 liters) of fuel for a whole race. That means engineers have to make sure the engines sip fuel like hummingbirds, even while blasting down straights at over 300 km/h.

The engines are so powerful that when a driver presses the accelerator, the car can go from 0 to 100 kilometers per hour (62 mph) in less than three seconds. That's about as quick as a rollercoaster launch, except this one doesn't stop after the first drop. In fact, an F1 car could beat most fighter jets off the line in the first 100 meters—at least until the plane's engines fully spool up!

Ever wondered why F1 engines are so loud? It's because they're pushing out hot air at an insane speed. The exhaust gases shoot out like a dragon breathing fire. In fact, before the rules changed in 2014, F1 cars were so noisy they could be heard miles away from the track. Some fans still miss that ear-splitting roar, describing it as "music for speed lovers."

Engines used to be even bigger. In the past, F1 cars sometimes had monstrous V12 engines—imagine 12 pistons firing at once! These beasts were heavy and thirsty, but they sounded like thunder rolling through the grandstands. Today's engines are smaller, hybridized, and more eco-friendly, but they're still just as powerful because of clever engineering, turbocharging, and battery power.

Here's a funny one: Drivers sweat so much in their cars (sometimes losing up to 3 liters of water in one race!) that the engine heat is partly to blame. The cockpit gets hotter than a pizza oven—seriously, over 50°C (122°F). Imagine racing at that speed while feeling like you're inside a sauna with a fire-breathing dragon under the hood. No wonder drivers sometimes collapse after the checkered flag.

The turbocharger in an F1 engine spins faster than almost anything else on Earth—up to 125,000 times per minute. That's so quick that if it were a merry-go-round, you wouldn't even see the horses—they'd be a blur of color. To stop the turbo from melting, it's cooled by a combination of special materials and advanced engineering that would look more at home on a spaceship.

Engines don't just need power; they need to be light. An F1 engine weighs only about 150 kilograms (330 pounds), lighter than three average 10-year-olds put together. The lighter the engine, the faster the car can accelerate and change direction. Engineers spend years shaving off grams of weight—sometimes even using 3D-printed parts that are hollow inside but just as strong.

Here's a gross-but-cool fact: Because of all the shaking and pressure, drivers sometimes feel their insides rattling during a race. One driver once joked that racing feels like "your organs are getting rearranged!" Not exactly comfy, but that's the price of speed.

F1 engineers love secrets. Teams spend millions designing engines that are just a little bit faster or more reliable than the competition's.

Sometimes, rival teams even accuse each other of sneaky tricks, like using fuel additives, bending the rules on oil burning, or programming unusual engine maps that give extra bursts of power.

And check this out: When an F1 car zooms past at full speed, it can suck in more air through its engine in one second than you could breathe in three minutes. That's a serious workout for the car's lungs! If you stood behind an F1 exhaust (don't try this!), the blast of hot air could knock you right over.

The fuel used in F1 isn't just powerful—it's clean. Modern rules require that at least 10% of the fuel comes from renewable sources, making it greener while still being lightning fast. Engineers are now experimenting with fully sustainable fuel for the future, hoping to make racing as eco-friendly as it is exciting.

An F1 car's engine produces about 1,000 horsepower. That's roughly the power of ten normal road cars combined, all packed into one lightweight racing machine. In fact, if you had that much power in a family car, just tapping the accelerator would send your groceries flying out the back window.

The hybrid system is just as important as the engine. F1 cars store extra energy when braking, then release it later for a boost. It's like having a video game turbo button in real life, and drivers time its use carefully—press it at the wrong moment and you've wasted your chance to overtake.

Temperature is everything. If the engine runs even a few degrees too hot, it could explode or seize up. That's why F1 cars have radiators hidden in their side pods to keep the engine cool—even though the car looks sleek and smooth. The balance is tricky: too much cooling slows the car down, too little cooling cooks the engine.

The engine doesn't last long. Each one only runs for about 1,500 kilometers before it needs replacing. That's the same distance a family car might cover in just two weeks. Imagine changing your engine that often—it would drive your mechanic crazy and drain your wallet fast. Teams only get a limited number of engines per season, so managing them is a strategy game all on its own.

Instead of a key, the engine is started with an external starter motor that plugs into the back of the car. If it stalls during a race, the driver can't just turn it back on—they need help from the pit crew. That's why a stall can be the end of a driver's race, even if the car itself isn't damaged.

Some engines in F1 history had bizarre designs. In the 1970s, there was even an experimental car with a fan attached to suck the car closer to the ground. It worked so well it was banned after a single race! Other unusual ideas included six-wheeled cars and rotary engines—but the strict modern rules keep things more standardized today.

Every part of the engine is designed with aerospace technology. Many materials used—like carbon composites and titanium alloys—are the same ones you'd find in rocket ships. That's why some fans say F1 cars are "planes without wings." The craftsmanship is so fine that some engine parts are thinner than a sheet of paper yet strong enough to handle enormous pressure.

At full throttle, the temperature of the gases leaving the exhaust can reach 1,000°C (1,832°F). That's hot enough to melt aluminum, and only special alloys can survive the inferno. If you could roast marshmallows over an F1 exhaust (not recommended unless you like instant charcoal), they'd be done in a blink.

Engineers spend endless hours on the tiniest details. Even the shape of the pistons and the pattern of fuel injection are tested in computer simulations before being built. One small mistake could cost a team millions of dollars, not to mention their shot at winning the championship.

The sound of an F1 engine is legendary. But here's a strange twist: because F1 cars run so efficiently, the sound you hear on TV isn't always the same as at the track. Microphones can't fully capture the ear-splitting scream of the engine—it's something fans say you need to hear in person at least once in your life. Some describe it as a howl, others as a buzzsaw, and some just say it makes your whole body vibrate.

Even the fuel lines are works of art. They're built to handle extreme pressure and heat without leaking a drop. If a fuel system fails, it's not just race-ending—it could be dangerous. That's why they're tested over and over before being allowed on track.

Finally, here's a fact most fans don't know: the engine in an F1 car is so tightly packed that mechanics sometimes need to remove half the car just to get to one part. Swapping an engine isn't like changing a tire—it's a marathon job that can take the entire night.

Engines in Formula 1 aren't built by just one person—they're designed by whole teams of scientists and engineers who work in secret labs. Some specialists focus only on the pistons, others only on the valves, and some just on making sure the parts don't melt under the pressure.

While the cars race on Sundays, the engines are tested thousands of times on dynamometers—giant machines that simulate the stress of racing without ever leaving the garage. These machines can run the engine flat-out for hours to check whether it will survive a real race.

Air is just as important as fuel. An F1 engine breathes in so much air during a single race that, if you trapped it all in balloons, you could fill an entire basketball court. The more air that goes in, the bigger the bang inside the cylinders, and the faster the car goes.

The rules of Formula 1 are strict: every team has to follow the same engine size limits. Right now, the maximum allowed is a 1.6-liter V6 turbo-hybrid. That might sound small—your family's car might even have a bigger engine! But thanks to advanced engineering, this tiny engine is a beast, producing nearly 1,000 horsepower.

The spark plugs inside an F1 engine are tougher than the ones in a normal car. They have to fire thousands of times every minute without missing a beat. Even one tiny spark failure could make the car slow down or stop completely.

Every second counts in racing, and that includes the time it takes for the engine to respond. When a driver presses the accelerator, the engine reacts instantly. Engineers call this "throttle response," and in F1 it's so sharp that the car seems to leap forward the moment the pedal is touched.

Sometimes, the cars run at full throttle for over 70% of a lap. That means the engine is screaming at maximum power almost constantly, with barely a chance to rest. In some races, like at Monza in Italy, engines are pushed harder than anywhere else—fans call it the "Temple of Speed" for a reason.

Engines don't just make the car go fast—they also help with braking. The hybrid system harvests energy when the driver slows down, using the resistance of the motor to recharge the battery. So in a way, slowing down actually helps the car go faster later.

Noise regulations have changed a lot over the years. Older fans remember the days of ear-splitting V10s and V12s, which were so loud they made the ground shake. Today's engines are quieter, but they're far more efficient, proving that speed doesn't always mean volume.

Even though every team follows the same rules, each engine has its own personality. Some are better at short bursts of speed, others are more reliable over long races. Drivers often talk about how their engine "feels," almost like athletes talking about the shoes they wear.

Here's a fact that will blow your mind: the oil inside an F1 engine isn't just for lubrication—it's part of the design. Engineers actually study how the oil moves around when the car is cornering at extreme angles, making sure it doesn't slosh away from where it's needed. Without that, the engine would seize up in seconds.

Engines also have to cope with wild weather. In hot races like Singapore, cooling is the biggest problem. In cold or rainy races, engineers need to adjust the fuel mixture and airflow to keep the engine running smoothly. That's why you'll often see mechanics huddled over laptops in the garage, constantly making changes even during the race.

The technology developed for F1 engines often ends up in road cars years later. Hybrid systems, lightweight alloys, and even fuel-saving tricks have all made their way into the cars you see on the highway. So, the engine secrets from the track eventually help everyday drivers.

One of the strangest things about F1 engines is that they rarely survive retirement. After a season, many engines are taken apart for study, displayed in museums, or simply scrapped. That means each one lives a short, intense life of speed before disappearing forever.

If you stood right next to an F1 car when it was revving at full power, the vibrations would rattle your chest like a giant drum. It's not just sound you hear—it's energy you feel. That's why fans wear ear protection, but still line up near the fences for that experience.

The engines aren't just powerful—they're smart. Built-in sensors measure temperature, pressure, and vibration hundreds of times per second. This data is sent back to the engineers, who watch live graphs and numbers during the race. If something looks wrong, they can warn the driver before disaster strikes.

Sometimes drivers are told to switch to a different "engine mode." This might mean saving fuel, protecting the engine, or unleashing maximum power for an overtake. It's like having multiple personalities in one machine. The driver changes it all with just a dial on the steering wheel.

Despite all the engineering, things can still go wrong. Explosions, smoke, or sudden failures still happen, reminding everyone that these machines are pushed right to the edge of what's possible. When an engine blows up, it's dramatic—sparks, flames, and a driver's race ruined in an instant.

Engines are so expensive that a single one can cost over $10 million to design, build, and test. And that doesn't even include the years of research behind it. That's why only the richest teams can compete in F1—without funding, there's no way to keep up with the technology race.

And here's one last jaw-dropper: if you added up the total power created by all 20 engines in a race, it would equal about 20,000 horsepower. That's the same as 200 Ferraris, all racing together on the same track. No wonder the ground shakes when the lights go out and the race begins.

Fun Challenge – Engine Detective!

Here are three curious engine "facts." One of them is fake. Can you guess which?

- Some F1 engines used to have 12 cylinders firing at once, making them sound like thunder.

- Drivers can restart their car by turning a key, just like your parents' car.

- The turbocharger in today's F1 engines spins faster than a jet engine turbine.

(Hint: F1 cars definitely don't have keys—pit crews use special starter motors instead!)

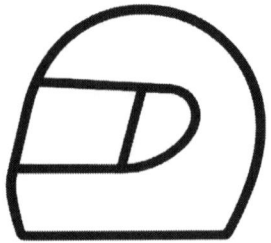

Chapter Two

Legends of the Track: Drivers Who Made History

Juan Manuel Fangio was nicknamed "El Maestro"—The Master—because he dominated the 1950s with five world championships. What made him legendary wasn't just his skill, but that he won with four different teams, proving he could adapt to any car.

Ayrton Senna is remembered not only for his three world titles but also for his fearless driving in the rain. His race at Donington Park in 1993 is often called the greatest opening lap in F1 history—he passed four cars in just a few corners, all while the track was soaked.

Michael Schumacher set records that stood for decades: seven world championships, 91 race wins, and a reputation for relentless focus. He was known for his extreme fitness training, treating himself more like an Olympic athlete than just a driver.

Lewis Hamilton tied and then beat Schumacher's win record. With seven world championships, over 100 race wins, and pole positions galore, he's not only one of the fastest but also one of the most consistent drivers in history.

Niki Lauda's story is one of the most famous comebacks in sports. In 1976, he nearly died in a fiery crash at the Nürburgring, suffering severe burns. Incredibly, he returned to racing just six weeks later, with bandages still on his face, and went on to win two more championships.

Sebastian Vettel once became the youngest world champion, starting a streak of four titles in a row from 2010 to 2013. Fans loved his cheeky personality—like when he named all his cars after women, such as "Kate's Dirty Sister."

Fernando Alonso earned the nickname "El Nano" and became Spain's first world champion. His back-to-back titles in 2005 and

2006 broke Schumacher's long streak, and his fiery determination still inspires fans even decades into his career.

Kimi Räikkönen, nicknamed "The Iceman," was famous for his cool, no-nonsense personality. He won the 2007 championship but also became a fan favorite because of his hilarious radio messages, like when he said, "Leave me alone, I know what I'm doing!"

Jim Clark, a Scottish driver from the 1960s, was considered so smooth behind the wheel that his car often looked slower than it really was—until the lap times proved otherwise. He won two championships and the legendary Indianapolis 500, showing his versatility.

Nigel Mansell, with his trademark mustache, was known for his aggressive style. His 1992 championship season was so dominant that he won nine out of 16 races, a record at the time. Fans loved his gutsy overtakes and his never-give-up spirit.

Jackie Stewart, nicknamed "The Flying Scot," not only won three championships but also fought hard for driver safety. Thanks to him, racing became much safer, with better helmets, medical facilities, and track design.

Alain Prost was called "The Professor" because of his clever, calculating style. He didn't always look the fastest, but his brainpower and strategy often won him the race. His battles with Ayrton Senna are still some of the most famous rivalries in F1 history.

James Hunt was as wild off the track as he was on it. Known for his playboy lifestyle and fiery personality, he won the 1976 championship in dramatic fashion, beating Niki Lauda in a season so crazy they made a Hollywood movie about it—*Rush*.

Stirling Moss never won a championship, but he's often called "the greatest driver never to win." He finished runner-up four times but is remembered for his incredible speed, fair play, and charisma.

Mika Häkkinen, known as "The Flying Finn," was the only driver who truly challenged Schumacher at his peak. With two championships, he earned respect for his daring overtakes and calm determination.

Daniel Ricciardo isn't just known for his race wins—he's also famous for his massive grin and "shoey" celebration, where he drinks champagne from his sweaty racing shoe. Gross? Yes. Memorable? Definitely.

Valtteri Bottas, another Finnish star, became famous for his lightning-quick starts. In 2017, he even jumped from third to first before the first corner, proving just how fast his reactions were.

Jenson Button became world champion in 2009 with a car that was almost unbeatable at the start of the season. Fans remember his smooth driving style and his incredible win at the 2011 Canadian Grand Prix, where he came from last place to first in a chaotic, rain-soaked race.

Graham Hill is still the only driver to win the "Triple Crown of Motorsport": the Monaco Grand Prix, the Indianapolis 500, and the 24 Hours of Le Mans. His legacy lives on through his son, Damon Hill, who also became world champion.

Damon Hill, following in his father's footsteps, won the 1996 championship and became the first son of a champion to also earn the title. Talk about family pressure handled well!

Record Breakers

Max Verstappen smashed record books when he became the youngest driver ever to start an F1 race at just 17 years old. Imagine juggling schoolwork while driving a car at 300 km/h! A few years later, he also became the youngest race winner, proving age was no barrier to speed.

Charles Leclerc, from Monaco, made history as the first driver from his tiny country to win an F1 race. Growing up just minutes away from the world-famous Monaco Grand Prix circuit, he used to watch the race from his apartment balcony. Now he's the one racing through those same streets.

George Russell earned the nickname "Mr. Saturday" for his knack of qualifying far higher than his car deserved. Even before he was in a front-running team, he shocked fans by putting slow cars at the sharp end of the grid.

Sergio "Checo" Pérez became the first Mexican driver in decades to win an F1 race. His comeback win in Bahrain 2020 was especially dramatic—he went from last place on the first lap to first at the checkered flag.

Drivers From Every Corner

Felipe Massa nearly became Brazil's next world champion. In 2008, he crossed the finish line in the final race as champion—only to lose the title seconds later when Lewis Hamilton overtook another car on the last corner of the last lap. Heartbreaking, but unforgettable.

Rubens Barrichello, also from Brazil, holds the record for the most race starts for many years, competing in over 300 races. His career spanned nearly two decades, making him one of the most experienced drivers ever.

Robert Kubica was the first Polish driver in F1 and won a race in 2008. After a near-fatal rally crash in 2011 that severely injured his arm, fans thought his career was over. But he made an inspiring return to F1 eight years later, proving his determination.

Pastor Maldonado was famous for his unpredictable style. He won one Grand Prix in 2012, shocking the racing world, but he also became known for constant crashes. Fans jokingly nicknamed him "Crashtor," but he was always entertaining.

Unbelievable Moments

Heinz-Harald Frentzen was once considered a rival to Michael Schumacher, and though he never won a championship, his surprising podiums with underdog teams made him a fan favorite.

Giancarlo Fisichella had one of the most emotional wins ever in Brazil in 2003. At first, officials mistakenly gave the victory to another driver, but days later the result was corrected. By then, he had already shipped his winner's trophy home—so it had to be mailed back to the rightful owner!

Jean Alesi, a French driver, only won one Grand Prix in his career, but it was legendary—Montreal, 1995. The crowd loved him so much they flooded the track after his win. His passion and style made him one of the most loved "one-win wonders."

David Coulthard, nicknamed "DC," spent years as a consistent front-runner with McLaren and Red Bull. Known for his sharp driving and equally sharp suits, he won 13 races and became a respected voice in commentary after retiring.

Funny & Quirky Legends

Takuma Sato was famous for his fearless overtakes. Sometimes they worked, sometimes they didn't—but they always got fans on their feet. His motto was "No attack, no chance."

Romain Grosjean became famous for both his crashes and his incredible escape in Bahrain 2020, where his car split in half and caught fire. Miraculously, he walked away with only burns on his hands—earning the nickname "The Phoenix."

Eddie Irvine, teammate to Michael Schumacher in the late 1990s, had a laid-back, cheeky personality. He once admitted he loved racing, but also enjoyed the parties just as much. Fans never knew what he'd say next in interviews.

Jarno Trulli was famous for what became known as the "Trulli Train." He was brilliant at qualifying high up the grid, but during races his car often lacked pace. Rivals stuck behind him for lap after lap, creating traffic jams on the track.

Inspiring Stories

Pierre Gasly scored one of the most emotional wins in modern F1 at Monza 2020. Driving for a smaller team, he beat the giants of the sport and became the first French winner in nearly 25 years. Fans celebrated it as a true underdog victory.

Esteban Ocon grew up in a modest background, with his parents even selling their house to support his racing dreams. Years later, he repaid that sacrifice by winning his first F1 race in Hungary 2021.

Kevin Magnussen gave Haas, one of the smallest teams on the grid, their first-ever pole position in 2022. His underdog story showed that even newer teams can have their moment in the spotlight.

More All-Time Greats

Stirling Moss may have been called "the best driver never to win a championship," but Sir Jackie Stewart carried a different crown: safety pioneer. Without Stewart's fierce campaigning, drivers today wouldn't have the strong helmets, HANS devices, and medical crews that keep them alive.

Emerson Fittipaldi, nicknamed "Emmo," became Brazil's first Formula 1 world champion in 1972. He was only 25 years old, making him the youngest champion at the time. His success paved the way for future Brazilian stars like Senna and Massa.

Nelson Piquet, another Brazilian legend, won three championships in the 1980s. Known for his cheeky humor and sharp mind, Piquet often played psychological games with rivals, proving that racing is as much about brains as bravery.

Rivalries & Friendships

Ayrton Senna and Alain Prost's rivalry defined the late 1980s. Their clashes—sometimes ending in crashes—became legendary. Yet, in a moving twist, Prost later carried Senna's coffin at his funeral, showing the deep respect behind their fiery battles.

Lewis Hamilton and Nico Rosberg grew up as friends and teammates, but their relationship cracked when they fought for the championship in 2016. The tension boiled over with wheel-to-wheel clashes, making their story one of the most dramatic teammate rivalries in history.

Sebastian Vettel and Mark Webber also clashed as Red Bull teammates. Their most famous moment was the "Multi-21" controversy, when Vettel ignored team orders and overtook Webber, leading to icy silence on the podium.

Courage Under Pressure

Gilles Villeneuve, father of Jacques Villeneuve, was adored for his fearless style. He would attack corners with wild slides, thrilling fans even when his car wasn't the fastest. Though he never won a championship, he's remembered as one of the most exciting drivers of all time.

Jacques Villeneuve, carrying on his father's legacy, became world champion in 1997. His bold moves and outspoken personality made him a star, especially when he stood up to Michael Schumacher in a title-deciding clash.

Riccardo Patrese's career lasted 17 years, and while he never won a championship, he earned six race victories and the respect of fans for his longevity. He once joked that he had "outlived" three generations of champions.

Quirky Driver Stories

Mark Webber famously crashed in a cycling accident before a race weekend and turned up with broken ribs. Instead of admitting it, he hid the injury from his team and still drove the car at high speed, wincing with every corner.

Kimi Räikkönen once skipped an official ceremony to go jet-skiing in a gorilla suit. This perfectly matched his "Iceman" personality—doing whatever he wanted, regardless of F1 formality. Fans loved his rebellious streak.

Jules Bianchi may have had a short career, but he left a lasting mark. He scored Marussia's first-ever points at Monaco in 2014, a miracle result for a small team. His bravery inspired a new generation of drivers.

Modern-Day Heroes

Carlos Sainz Jr., nicknamed "Smooth Operator" thanks to his radio catchphrase, became Spain's next big hope after Fernando Alonso. His consistency and clever driving earned him podiums and admiration worldwide.

Lando Norris quickly became a fan favorite for his humor and video game streaming off the track. On the track, he proved his skill with daring overtakes and podium finishes, showing he's part of the new generation ready to dominate.

Fun Activity – Build Your Dream Team

Imagine you're the boss of a new F1 team. You get to pick **three drivers from history** to race for you. Who would you choose, and why?

Here are some categories to help you decide:

Speed Demon: Who's the fastest driver you've read about?

Tactical Genius: Who's the cleverest strategist on track?

Showman: Who would make the fans cheer the loudest?

Write down your dream team and explain your choices. There's no wrong answer—just have fun mixing legends from different eras.

Chapter Three

Pit Stop Madness: The Fastest Crew on Earth

The Need for Speed

A modern Formula 1 pit stop can take less than two seconds. That's faster than you can say "Go, go, go!" In 2019, Red Bull set the all-time record with a stop that lasted just 1.82 seconds. Blink, and you'd miss it.

It wasn't always this way. In the early days of F1, pit stops often lasted over a minute. Mechanics had to use hand tools, and drivers sometimes even lit a cigarette while waiting for refueling and tire changes. Today, even half a second too long can cost a race win.

Pit crews train like athletes. Many members come from backgrounds in sports like rugby or weightlifting. They spend hours practicing their moves, doing strength training, and even rehearsing with stopwatches to shave off tiny fractions of a second.

Every person in a pit stop has a specific role. One mechanic works the front-right wheel, another the rear-left, others handle the jacks, and one signals the driver when to go. With around 20 people surrounding the car, it looks like a perfectly choreographed dance.

The wheel guns used in F1 aren't normal tools. They're specially designed pneumatic guns that spin at 10,000 revolutions per minute. That's faster than a dentist's drill, but instead of teeth, they're zapping on and off massive wheel nuts.

Speaking of nuts, each F1 wheel is held on by just one giant nut in the center. Unlike road cars with five small bolts, F1's single-nut design means faster changes—but if it's cross-threaded or stuck, the car might be stuck in the garage.

High-Tech Tools & Tricks

The jacks used to lift the cars are lightweight but incredibly strong. The front jackman has to time his move perfectly—lifting the car the instant it stops. If he's too slow, the whole stop falls apart.

Rear jacks have a "quick release" system, so as soon as the tires are changed, the car can drop instantly. Sometimes, mechanics even practice blindfolded so they know their motions by feel alone.

The lollipop man, named after the paddle used to signal the driver, has mostly been replaced by traffic lights. But in the past, this role was crucial—one wrong move and the driver could be released into oncoming traffic or before the wheels were secure.

Teams use practice cars in their garages just for pit stop rehearsals. Crews will sometimes repeat the same move 50 or 60 times in one afternoon until it becomes second nature. Drivers also practice stopping precisely on their marks—because being even 10 centimeters off can slow everything down.

Chaos & Comedy

Not every pit stop goes smoothly. In 1999, Ferrari's Eddie Irvine had a pit stop disaster when his team accidentally fitted three tires of one type and one tire of another. The mix-up cost him the race and became one of the most infamous pit stop blunders.

Another classic fail happened to Jenson Button in 2011, when a wheel nut rolled under his car during the stop. The crew scrambled to find it, wasting precious seconds, while Button sat helplessly in the cockpit.

Sometimes, pit crews forget to bring the tires out at all. In 2020, Mercedes left George Russell waiting for nearly half a minute because the wrong set of tires was prepared. He'd been leading the race but dropped back instantly.

And here's a funny one: in 2008, Ferrari accidentally released Felipe Massa with the fuel hose still attached. He sped down the pit lane dragging it behind him like a tail, until he stopped and the crew had to chase after him to pull it out.

Pit lane mistakes aren't always small. Unsafe releases—where two cars nearly collide in the pits—have led to time penalties and tense radio messages. Drivers trust their crews completely, but one miscommunication can turn everything upside down.

The Human Element

Pit crews don't just change wheels. They also swap front wings, adjust aerodynamics, and clean out debris stuck in the car's bodywork. Every second counts, so they've trained to do it all in lightning-fast motions.

The physical demands are intense. Crew members practice holding heavy tires that weigh over 20 kilograms each and running with them at full speed. Their reaction times are trained using light boards, just like athletes preparing for Olympic sprints.

Being in the pit crew isn't glamorous. Mechanics work long hours in the garage, often sleeping on the floor during race weekends. But when the car wins thanks to a perfect pit stop, they get to celebrate just like the driver.

How Pit Stops Evolved

In the earliest days of racing, pit stops were glacial compared to today. Mechanics sometimes had to hammer on wheels, pour fuel from jugs, and patch up broken parts—all while the driver waited patiently. Some stops lasted over ten minutes, which would be unthinkable now.

By the 1980s and 1990s, pit stops were down to around seven seconds. At the time, that was jaw-droppingly fast. Broadcasters even ran slow-motion replays so fans could appreciate the choreography. Today, anything slower than three seconds is considered sloppy.

Refueling was part of pit stops until 2010. Drivers would stop not just for tires, but also to gulp down liters of fuel. The practice was banned for safety reasons, as fires were far too common. One famous example was Jos Verstappen's pit stop in 1994, when his car burst into flames during refueling. Luckily, he escaped with only minor burns.

Pit Lane Drama

Pit crews are under enormous pressure, and mistakes can be dramatic. In 2013, Mark Webber's Red Bull car was released from the pits with a loose wheel that bounced down the lane like a cannonball. It even hit a cameraman, leading to stricter safety rules.

In 2018, Kimi Räikkönen's Ferrari crew accidentally gave him the green light too early. One mechanic's leg was broken when the car drove away before the wheel was secured. The team apologized, and the mechanic made a full recovery—but it showed just how risky life in the pits can be.

Sometimes the drivers themselves cause chaos. In 1991, Nigel Mansell accidentally stalled his engine in the pit lane, holding up everyone behind him. The crew had to frantically restart his car while rival drivers laughed as they zipped past.

Science of Perfection

The choreography of a modern pit stop is mind-blowing. A team of about 20 mechanics surround the car in a blur of motion, each person trained for a single movement. A front jackman lifts, four wheel gunners blast the nuts off, four mechanics pull the tires away, and four slam on the new ones. The whole sequence is over in less time than it takes you to tie one shoe.

Every movement is analyzed on video afterward. Teams study slow-motion replays to see if a mechanic hesitated, or if a wheel gun slipped for a fraction of a second. A single blink of hesitation can cost an entire race.

Even the way crews hold their tools is planned. For instance, wheel gunners practice holding their air guns at a precise angle so they connect with the nut perfectly every time. It's like a racing pit stop version of martial arts.

Training Like Athletes

Many pit crew members were once athletes. Rugby players, sprinters, and even weightlifters are recruited because of their strength, speed, and coordination. Being able to squat with a 20-kilo tire and sprint five meters in under a second is part of the job description.

Crews use reaction-training boards—flashing lights that they have to tap as quickly as possible—to keep their reflexes razor sharp. They even practice with virtual simulations, using high-tech computer programs to predict the fastest possible routines.

Pit stops are also dangerous for the crew's bodies. Mechanics often suffer back problems, knee injuries, and strained muscles from constantly lifting heavy tires and working at awkward angles. Just like drivers, they need physiotherapists to keep them in shape.

Wild & Weird Pit Stop Stories

In 2006, during the Monaco Grand Prix, a car was released right in front of another in the cramped pit lane. Both drivers ended up stuck side by side like cars in a traffic jam. The mechanics had to physically lift one car backward to make space.

Another bizarre moment came when an entire pit crew once wore wigs and sunglasses during a practice session just to prank their driver. He pulled into the pit box to find what looked like a team of Elvis impersonators changing his tires. Even in high-pressure racing, humor sneaks in.

And in one unusual case, a driver stopped in the wrong pit box entirely, confusing the rival team's crew. Imagine trying to change tires for a car that doesn't even belong to you!

The Fastest Hands in Racing

Red Bull isn't the only team with lightning pit stops. Williams once held the record, and McLaren, Ferrari, and Mercedes have all had runs where their stops consistently broke the two-second barrier. The battle for "fastest crew" is almost as competitive as the drivers' championship itself.

Teams measure pit stop times down to thousandths of a second. A difference of 0.2 seconds might sound tiny, but on track it could be the gap between first and fifth place once the pit cycle is complete.

Perfect Precision

The car itself is designed to help the pit crew. Wheel nuts are engineered so they can't roll away, and the wheels are shaped to slot perfectly into place. Even the driver's job is crucial: they must stop on an exact mark, no more than a few centimeters wide, otherwise the crew can't reach the car properly.

Drivers practice hitting that mark hundreds of times in simulators. A sloppy stop isn't just the crew's fault—if the driver brakes too early or late, the entire rhythm is ruined.

High-Stakes Strategy

Pit stops aren't just about speed—they're about timing. Teams use powerful computers to decide the perfect lap for a pit stop. A stop that's one lap too late can mean being stuck behind slower traffic, while a perfectly timed stop can leapfrog a rival.

Sometimes teams gamble by "double stacking"—pitting both their cars one after the other. If executed perfectly, it's a huge time saver. If not, the second driver sits waiting in line while seconds tick away.

When It Goes Wrong

Loose wheels are one of the most dreaded mistakes. If a car leaves the pit box without a wheel fully attached, the driver must stop immediately. Continuing on could damage the car or cause an accident. Teams practice over and over to avoid this nightmare.

In 1994, a wheel flew off a car in the pits and bounced into the grandstands, injuring several fans. Since then, rules and barriers have been tightened to protect spectators from runaway tires.

Pit Stop Heroes

Not all pit stops are about tires. In 2008, during the Belgian Grand Prix, a pit crew actually used tape to fix a broken wing on Kimi Räikkönen's car in just a few seconds. It wasn't pretty, but it got him back out on track.

In another race, a team changed not only all four tires but also the driver's steering wheel in under 10 seconds. The wheel is like a computer with dozens of buttons, yet the crew swapped it faster than most people can plug in a video game controller.

Training Never Ends

During a typical race weekend, pit crews might practice 50 to 75 stops just to stay sharp. Some teams even bring full-size dummy cars to every race, purely for training.

The physical training is grueling. Crews work on sprinting, balance, and explosive strength so they can lift, pull, and move in perfect harmony. They may not drive the car, but they're athletes in their own right.

The Unsung Champions

When a driver wins, the headlines usually focus on their speed. But drivers are the first to admit that without their pit crews, victories wouldn't happen. A perfect pit stop can turn a close race into a runaway win, and a bad one can crush championship hopes.

Fans sometimes cheer louder for a two-second pit stop replay than for a daring overtake on track. That's how important the crews have become to the theater of Formula 1.

Activity – Pit Stop Challenge

Imagine you're the chief of your very own F1 pit crew. You have 20 people to assign, and each one has to do exactly the right job. Here are the roles:

- Four wheel gunners

- Four tire removers

- Four tire fitters

- One front jackman

- One rear jackman

- Two stabilizers to hold the car steady

- One lollipop man or traffic light controller

- Two spare mechanics in case something goes wrong

- One team leader to signal the all-clear

Your challenge: write down how you would position them around the car. Would you put your fastest people on the front wheels? Who gets the hardest job? Explain your choices, and see if you can design the ultimate two-second pit stop.

Chapter Four

Tracks Around the World: From Monaco to Monza

The Jewel of Monaco

The Monaco Grand Prix is the most glamorous race in the world. Drivers race through narrow city streets past million-dollar yachts, luxury casinos, and balconies filled with cheering fans. Winning here is so special that even champions call it the highlight of their career.

Overtaking in Monaco is nearly impossible. The streets are so tight that one small mistake can block the entire track. In 2004, a crash completely clogged the tunnel, creating one of the most bizarre traffic jams in F1 history.

The famous Fairmont Hairpin in Monaco is the slowest corner in Formula 1. Cars crawl through it at just 50 km/h—slower than your parents driving through a supermarket parking lot. But because of the car's sharp steering, it's still one of the hardest corners to master.

Monza – The Temple of Speed

Italy's Monza circuit has been around since 1922, making it one of the oldest tracks still in use. It's called "The Temple of Speed" because drivers are at full throttle for nearly 80% of the lap.

Monza once featured enormous banked corners, like a giant oval racetrack. Drivers flew sideways up the banking, sometimes so high they looked like they were driving on a wall. The banking still exists today as a ghostly reminder of old-school danger.

The Italian fans, called *tifosi*, turn Monza into a sea of red every September. After Ferrari wins, thousands of fans climb over fences and flood the track in a celebration that looks more like a rock concert than a race.

Silverstone – Where It All Began

The very first Formula 1 World Championship race took place at Silverstone in England in 1950. Back then, the track was built on an old World War II airfield, with hay bales marking the corners and even sheep grazing nearby.

Silverstone is still on the calendar today and has grown into one of the fastest and most technical tracks. Corners like Copse and Maggotts-Becketts are so quick that drivers' necks feel like they're being yanked sideways by an invisible giant.

Suzuka – The Figure Eight

Japan's Suzuka circuit is the only track in F1 shaped like a figure eight. Drivers race over a bridge and under it, making it one of the most unique layouts in motorsport.

The Suzuka "S-Curves" are a series of left-right corners that test a driver's rhythm and precision. Even the smallest mistake here can ruin an entire lap. Fans say watching cars glide through the S-Curves is like watching a perfectly choreographed dance.

Spa-Francorchamps – The Rollercoaster

Belgium's Spa-Francorchamps is famous for its hills and forests. The track is so big it can be raining on one side while sunny on the other, making tire choices a nightmare.

The Eau Rouge corner at Spa is one of the most legendary in racing. Drivers blast downhill, then sweep uphill so fast that their stomachs flip like they're on a rollercoaster. Some say it takes more courage to go flat-out here than anywhere else on Earth.

Nürburgring – The Green Hell

The Nürburgring in Germany was so long and so dangerous that Jackie Stewart nicknamed it "The Green Hell." At over 22 kilometers per lap, it twisted through forests, with more than 150 corners. Drivers had to memorize every bend, often saying it felt like fighting a wild dragon.

In 1976, Niki Lauda had his fiery crash here, which changed safety rules forever. Today, F1 only races on the shorter modern Nürburgring track, but the legend of the old circuit lives on.

Circuit de la Sarthe – Borrowed Roads

While not a regular F1 track, Le Mans' Circuit de la Sarthe is famous because parts of it are just public roads closed off once a year. Formula 1 drivers who also competed there often said it felt surreal to race past streetlights and signposts at full throttle.

Interlagos – Samba Spirit

Brazil's Interlagos circuit sits between two lakes, which is how it got its name. The track is short, twisty, and hilly, making it one of the most exciting to watch. The atmosphere is electric, with Brazilian fans drumming, waving flags, and dancing in the stands.

This track has hosted some of the most dramatic championship deciders, including the 2008 finale when Lewis Hamilton clinched the title on the very last corner of the last lap.

Circuit Gilles Villeneuve – Wall of Champions

Canada's Montreal circuit is named after Gilles Villeneuve, a driver loved for his fearless style. But the track has a scary landmark—the "Wall of Champions." Many world champions, including Damon Hill, Michael Schumacher, and Jacques Villeneuve, have crashed into it.

The track also has long straights leading to tight chicanes, which means lots of overtaking and plenty of drama.

Singapore – Racing Under the Lights

Singapore's Grand Prix was the first ever held entirely at night. The track is lit by thousands of floodlights, making it brighter than daylight. Racing through a glowing city skyline is a spectacle like no other.

The humidity, though, is brutal. Drivers often lose up to three kilograms in sweat during the race, and by the finish line, they look like they've just stepped out of a swimming pool.

Circuit of the Americas – Stars and Stripes

In Austin, Texas, the Circuit of the Americas (COTA) brings American flair to Formula 1. The track has a steep uphill climb into the first corner, where cars slow from 320 km/h to a sharp hairpin. Fans call it one of the best overtaking spots in modern F1.

COTA also borrows some of its design from famous tracks around the world, like Silverstone's sweeping S-Curves. It's like a "greatest hits" circuit rolled into one.

Hungaroring – Monaco Without Walls

Hungary's Hungaroring is nicknamed "Monaco without the walls" because it's narrow and twisty, making overtaking incredibly difficult. Races here often turn into long battles of strategy instead of raw speed.

The hot summer weather also makes it tough on both cars and drivers, with cockpit temperatures soaring. Victories here are celebrated as hard-earned triumphs.

Mexico City – Stadium Roar

The Autódromo Hermanos Rodríguez in Mexico has one of the most unique features in F1: cars race straight through a giant baseball stadium. Fans pack the stands, and when a Mexican driver like Sergio Pérez passes through, the roar is deafening.

The high altitude makes engines and drivers struggle, since the air is thinner. Teams must adjust everything, from turbochargers to cooling systems, just to keep the cars running.

Baku – The City of Contrasts

The Baku City Circuit in Azerbaijan combines wide, flat straights with tight medieval corners. The track even squeezes past a centuries-old castle wall. Drivers hit over 350 km/h on the main straight, then crawl through narrow sections barely wider than the car itself.

It's a track of chaos—safety cars and crashes are common, and races here often flip upside down in an instant.

Albert Park – Party in the Park

Melbourne's Albert Park is a street circuit that weaves around a lake and public park. For one weekend a year, roads used by joggers and cyclists turn into a roaring F1 track.

As the traditional season opener for many years, the race has a carnival atmosphere, with fans picnicking in the grass while the cars thunder past.

Imola – History and Heartbreak

Imola in Italy is steeped in both triumph and tragedy. It's where Ayrton Senna lost his life in 1994, leading to sweeping safety changes in F1. But it's also remembered for thrilling battles and passionate Italian fans who still pack the grandstands today.

The circuit winds through the countryside, giving it an old-school feel that many drivers still love.

Zandvoort – Orange Madness

The Dutch Grand Prix at Zandvoort has returned to the calendar with roaring success. What makes it stand out isn't just the track's banked corners, but the sea of orange-clad Max Verstappen fans. The grandstands erupt like a volcano every time their hero passes by.

The banked final corner is so steep it feels more like an amusement park ride than a racetrack bend, challenging drivers to hold their line at high speeds.

Bahrain – Desert Showpiece

Bahrain made history in 2004 as the first Formula 1 race in the Middle East. The track sits in the middle of the desert, with sandstorms sometimes blowing across the tarmac.

Night races under the desert floodlights create a dramatic glow, with sparks flying from the cars' underbodies. It's one of the most visually stunning events of the season.

Yas Marina – Modern Marvel

Abu Dhabi's Yas Marina Circuit takes racing luxury to the next level. The track runs past a harbor filled with superyachts, under a glowing five-star hotel that changes color during the race.

The pit lane has a one-of-a-kind feature: cars exit through a tunnel that dips under the track. It looks more like a video game than real life.

Sepang — The Malaysian Giant

The Sepang International Circuit in Malaysia was known for its massive grandstands and unpredictable tropical weather. It could be blazing hot one minute and pouring rain the next, turning the track into a water park.

Even though it's no longer on the F1 calendar, fans remember its sweeping corners and sudden storms that shook up races.

Shanghai — The Dragon's Eye

China's Shanghai International Circuit is shaped like the character "shang," meaning "above." Its long back straight stretches for 1.2 kilometers, one of the longest in Formula 1. Cars hit blistering speeds before slamming on the brakes for a sharp hairpin.

The grandstands are designed to look like giant lotus flowers, adding a cultural touch to the futuristic layout.

Barcelona — Testing Ground

The Circuit de Barcelona-Catalunya in Spain is used heavily for pre-season testing. Drivers often joke that they know every bump in the track by heart because they've driven so many laps there.

While races here aren't always the most dramatic, engineers love the variety of corners, which reveal a car's strengths and weaknesses.

Las Vegas – Lights and Luck

F1 is returning to Las Vegas with a street circuit that races down the famous Strip. Imagine neon lights, casinos, and fast cars all blending into one electric spectacle.

The last time F1 raced in Vegas, in the 1980s, the track was literally built in a hotel parking lot! The modern version promises to be much more glamorous.

Adelaide – Party Circuit

Before Melbourne, Australia's F1 race was held in Adelaide. The circuit was famous for its lively crowds and dramatic season finales.

In 1986, three championship contenders crashed or broke down, handing the title to Alain Prost in one of the most shocking twists in F1 history. Fans still talk about Adelaide as one of the best street races ever.

Kyalami – African Spirit

South Africa's Kyalami circuit once hosted F1 races, bringing the sport to the African continent. The track had fast, flowing corners and passionate fans who created a festival-like atmosphere.

Though it hasn't been on the calendar for decades, there are talks of F1 returning, which would make it a truly global championship again.

Indianapolis – The Oval Meets F1

The United States Grand Prix once took place at Indianapolis, where part of the race used the famous oval banking. Mixing the oval with twisty infield corners created a unique layout.

In 2005, the race turned into chaos when tire safety concerns left only six cars competing. It was one of the strangest F1 events ever held.

Paul Ricard – Stripes of Confusion

France's Paul Ricard Circuit is instantly recognizable thanks to its red and blue striped run-off areas. These colorful stripes aren't just decoration—they're made of special abrasive material designed to slow cars down if they go off track.

The wide layout and rainbow colors sometimes make it tricky for TV viewers to even tell where the actual track ends!

Activity – Track Builder Challenge

Imagine you're designing your very own Formula 1 circuit. You can borrow features from any of the tracks you've read about:

Will it have Monaco's tight streets or Monza's flat-out straights?

Will it include a banked corner like Zandvoort or a night-race glow like Singapore?

Will fans watch from grandstands, beaches, or even boats?

Draw your dream layout on paper, give it a name, and write a short description of why your circuit would be the most exciting on the calendar.

Chapter Five

Epic Crashes & Close Calls: The Wild Side of Racing

Formula 1 crashes look violent, but the whole car is actually designed to break apart like a giant Lego set. The nose, sides, and wings shatter in a crash because they're supposed to. That way, the energy is spread out instead of smashing into the driver's body. The part that never breaks is the "survival cell"—a carbon fiber capsule where the driver sits, strong enough to resist more force than a military tank.

The 1955 disaster at Le Mans wasn't a Formula 1 race, but it changed the sport forever. A crash killed more than 80 spectators, and the shock made F1 rethink everything: trackside barriers, catch fencing, pit lane safety, and crowd placement. It was the moment racing realized thrills weren't worth lives, and from that point on, safety slowly became a science.

One of the most famous crash stories belongs to Niki Lauda at the Nürburgring in 1976. His car burst into flames and he was trapped inside, suffering burns that scarred him for life. Fellow drivers pulled him out in time. Astonishingly, Lauda was back racing six weeks later, his wounds still healing. More important than his comeback was the impact on the sport—fireproof suits, marshals with faster response times, and serious debates about whether the Nürburgring was too dangerous.

Rain makes every race a lottery, and crashes become almost inevitable. In 1998 at Spa, rain on the first lap created a wall of spray so thick drivers couldn't see. One car spun, and suddenly more than a dozen cars piled up in one of the biggest wrecks ever seen in Formula 1. No one was badly hurt, which was lucky, but it showed how easily water can turn a race into a demolition derby.

Robert Kubica's crash in Canada 2007 looked like something no one could survive. His car slammed the wall at over 230 km/h, flipped, and scattered into a thousand pieces. Fans braced for the worst. Amazingly, Kubica only sprained his ankle and had a sore hand. The car's design saved him—the cockpit cell stayed intact, and modern TecPro barriers absorbed most of the force. One year later, he returned to the same track and won.

The halo device is now one of the biggest lifesavers in F1. It's a titanium wishbone over the cockpit that can carry the weight of a London bus. At first, some fans hated the look. But then Charles Leclerc, Lewis Hamilton, and Zhou Guanyu all walked away from accidents where cars or wheels landed on top of them. Without the halo, those stories would have ended very differently.

Romain Grosjean's crash in Bahrain 2020 was a nightmare. His car pierced a barrier, split in half, and exploded in flames. He was trapped for 28 seconds before climbing out, fire around him. His gloves melted, and his hands were badly burned, but he survived. Engineers studied every detail of that crash—fuel lines, gloves, barriers—and upgraded systems again. Grosjean called himself "The Phoenix," and fans agreed.

Flying wheels were once a terrifying danger. They used to bounce across tracks like giant bowling balls when cars crashed. That changed with the invention of wheel tethers—fibers hidden in the suspension that are strong enough to hold a car even if the wheel snaps off. They don't always work perfectly, but they've massively reduced the chance of loose wheels hitting other drivers or fans.

Medical response is another reason more crashes are survivable today. At every race, a fully equipped medical car follows the grid on the opening lap. Inside are a doctor and equipment ready for anything. Trackside medical centers are as advanced as hospital emergency rooms, and helicopters are always waiting if a transfer is needed. When Grosjean's crash happened, medics were at his side in seconds—that speed saved him.

Crashes in the pit lane can be just as scary. In 2008, Felipe Massa left his pit box still attached to a giant fuel hose, dragging it behind him like a tail. In 2013, Mark Webber's car left the pits with a loose wheel that bounced down the lane and hit a cameraman. Incidents like these led to stricter pit rules, new technology, and penalties for unsafe releases. Even a two-second stop carries real danger if things go wrong.

And then there are the smaller, stranger details—the ones drivers laugh about later. After gravel-trap excursions, they shake pebbles from their boots. Mechanics joke about "black boogers" after a weekend surrounded by carbon brake dust. Gross? Maybe. But it's also why modern garages have heavy ventilation, and why mechanics wear gloves and masks. Even the silly side of a crash connects to safety science.

Not all dramatic crashes happen at high speed. Sometimes it's the unexpected, slow-speed accidents that shock the most. In 2012 at Monaco, Pastor Maldonado clipped another car at a corner and triggered a pile-up. Because Monaco's streets are so narrow, cars blocked each other like a traffic jam, and drivers had to sit helplessly until marshals cleared the mess. It was a reminder that danger isn't just in the straights—tight corners can cause chaos too.

Spectacular flips have dotted F1 history. In 1999 at Monza, Alex Zanardi's car was launched into the air after contact, spiraling before landing back on the track. More recently, in 2010, Mark Webber hit the back of another car in Valencia and flew so high that fans gasped he looked like a fighter jet. Thankfully, he walked away with nothing worse than bruises—proof that modern cockpits are built like fortresses.

One of the strangest incidents came in 2007 at the Canadian Grand Prix, when a groundhog wandered onto the track. Drivers dodged it at full speed, and while it wasn't a crash between cars, it showed how even wildlife can cause close calls. Animals like birds and dogs have also found their way onto circuits over the years, making races unpredictable in more ways than one.

Crashes in rain are especially tricky because cars can spin even in a straight line. At the 2007 Japanese Grand Prix, visibility was so bad that drivers radioed they couldn't see the car directly in front of them. Cars slid off the track left and right, turning the race into survival on a soaked rollercoaster. Engineers have since studied spray patterns and even tested mudguard-style devices, though they haven't yet been adopted.

Sometimes drivers make mistakes under pressure in the pits, creating dangerous moments. In 1991, Nigel Mansell stalled his car during a stop and blocked the entire pit lane. Mechanics scrambled to restart him while other drivers queued behind, shaking their heads. Pit lane rules are now stricter, with time penalties for unsafe or delayed releases to prevent pile-ups.

Bravery is often tested at high-speed corners. Eau Rouge at Spa is legendary not only for thrilling laps but also for horrifying crashes when drivers lose grip mid-curve. In 1999, Ricardo Zonta and Alex Wurz both had massive accidents there in practice but escaped injury. It reinforced why gravel traps, barriers, and constant safety tweaks are crucial on tracks where drivers go full-throttle uphill into the unknown.

Safety gear has evolved alongside cars. Helmets once looked like leather caps, offering little protection. Today's helmets use carbon fiber, kevlar, and fireproof lining. During Felipe Massa's 2009 Hungary accident, when a spring from another car struck his helmet at 260 km/h, the impact could have been fatal. The visor and shell absorbed the blow, saving his life. It led to even stricter helmet safety tests.

Brakes are another hidden factor in crashes. F1 brakes get so hot they glow red and reach temperatures above 1,000°C. If a brake fails, stopping power vanishes instantly. In 2001, Luciano Burti's car plowed straight into a barrier after a brake issue, but his HANS device—straps connecting helmet to shoulders—prevented serious neck injury. It's a reminder that even invisible parts can make or break safety.

Close calls don't just happen to drivers. Marshals, the orange-suited volunteers trackside, have been in danger too. In 2001 at Melbourne, a marshal was tragically struck by flying debris, which pushed the sport to add more protective fencing and stricter debris tether rules. These days, marshals are trained like professionals, often risking their safety to keep races running smoothly.

Crashes can be dramatic even after the finish line. In 1991, Nigel Mansell tried to wave to the crowd on his victory lap but stalled his car. He had to push it back while marshals laughed and fans cheered. It wasn't dangerous, but it showed how fragile F1 cars can be when handled the wrong way—even by the winners.

And of course, there are the moments drivers cause accidents all by themselves with overconfidence. Romain Grosjean's first-lap crash at Spa in 2012 wiped out several cars and earned him a one-race ban. He admitted later he had been too aggressive. It was a lesson for every young driver: skill matters, but so does restraint, especially when 20 cars are crammed into the first corner.

Formula 1 cars sometimes look indestructible, but even tiny mistakes at high speeds can cause huge accidents. In 1994, Jos Verstappen's car caught fire during a refueling stop when fuel sprayed onto the hot exhaust. The fireball lasted only a few seconds before marshals put it out, but it showed how dangerous mid-race refueling could be. That's one reason why refueling is no longer allowed today.

Even safety cars have seen their share of trouble. At the 1999 Canadian Grand Prix, the safety car itself crashed into the wall while leading the pack. Drivers behind couldn't believe it—the very car meant to keep them safe had spun off! Ever since, safety car drivers are tested and trained just as much as the racers they control.

Sometimes it isn't fire or speed that creates drama but confusion. In 2007 at Fuji Speedway, a driver tried to rejoin the race from the pit exit but accidentally turned the wrong way and found himself facing the field head-on. Officials black-flagged him instantly, but it showed how high-pressure moments can scramble even elite drivers' instincts.

The Indianapolis Grand Prix in 2005 created one of F1's strangest "crash-free" disasters. Tire safety concerns meant that only six cars could start, while the rest pulled out before the race. Fans booed, teams argued, and the whole event went down as one of the sport's most embarrassing close calls. It wasn't a pile-up, but it showed how safety can end a race before it begins.

Drivers sometimes crash after the race is over. In 2009, Jarno Trulli and Adrian Sutil collided on the cool-down lap, then argued trackside while their broken cars sat as props. Fans couldn't believe it—most accidents happen when drivers are fighting for points, not after the checkered flag.

One of the scariest airborne crashes in recent memory happened at Silverstone in 2022. Rookie Zhou Guanyu's car flipped upside down at the start and skidded along the track, sparks flying as the roll hoop dug into the asphalt. The car then slid into fencing near the crowd. Thanks to the halo and the survival cell, Zhou walked away unharmed, proving once again how far safety has come.

Accidents don't always look spectacular but can be just as dangerous. Brake failures, steering malfunctions, or suspension collapses have sent cars straight into barriers with no warning. In 2018, Lance Stroll's front suspension snapped at high speed in practice, and he hit the wall hard. Engineers studied the failure and reinforced designs to make sure it wouldn't happen again.

Big crashes can even be caused by small animals. In 2016, during practice in Russia, a dog ran across the track just as cars approached at full speed. Marshals had to scramble to herd it away before disaster struck. Birds, groundhogs, and even lizards have made cameo appearances in F1 races, each one creating a risky close call.

The G-forces in accidents are staggering. In 2003, Fernando Alonso crashed in Brazil and experienced forces over 35 g. To imagine that, think of your body suddenly feeling 35 times heavier in a fraction of a second. That's why drivers train their necks and cores so heavily—and why crash testing is done at terrifying speeds.

Sometimes crashes turn into funny moments once the danger has passed. In 2008, David Coulthard crashed out of his final race, then hitchhiked back to the pits by climbing onto another driver's car during the lap back to the garage. Fans laughed at the sight of a legendary driver "catching a ride" in full gear.

And then there are the moments that remind everyone why safety is always evolving. Every serious accident—Lauda's fire, Senna's tragedy, Grosjean's fireball, Zhou's flip—leads to new innovations. Stronger cockpits, better helmets, safer barriers, quicker marshals, and smarter rules all trace back to hard lessons. Formula 1 learns from every close call, making each crash a step toward a safer future.

Activity – Crash Science Detective

Pick one of these crash factors and explain why you think it's the most important in keeping drivers safe:

- Barriers (TecPro, tire walls, SAFER barriers)

- The survival cell (the cockpit "capsule")

- The halo device

- Fireproof clothing

- The medical car and marshals

Write down your choice, then argue your case to a friend or family member. Do they agree with you, or do they think another safety feature is more important?

Chapter Six

Weird, Gross, and Funny F1 Facts

Sweaty, Smelly, and Just Plain Gross

Formula 1 drivers sweat so much during a race that they often lose between 2–3 kilograms of body weight in under two hours. That's the same as lugging around a giant watermelon, then magically making it vanish through sweat. In Singapore, where heat and humidity combine, drivers describe stepping out of the car as feeling like they just swam laps in their racing suits.

And speaking of suits, they're not exactly bathroom-friendly. Once the helmet is strapped on, there are no pit stops for the bladder. Many drivers admit they've had to pee in their suits during long races. Lewis Hamilton once confessed to it in interviews, and Nico Rosberg said that sometimes drivers don't want to admit it but... it happens. Kids laugh, but drivers just shrug—it's all part of the job.

Helmets also get their share of grossness. When a driver is dehydrated, saliva gets sticky, and instead of swallowing, some drivers spit inside the helmet. This keeps their mouths clear, even if it means sloshing around in their visors. It's not glamorous, but vision comes first—better a spit-filled helmet than a foggy visor at 300 km/h.

Strange Superstitions

Many top drivers rely on quirky rituals to keep calm. Michael Schumacher always got into his car from the same side, no matter what. Even if mechanics had to awkwardly make space, he wouldn't budge from his routine.

Ayrton Senna often prayed in the car before every race, and some said he had a ritual of closing his eyes on the grid to center himself.

Sebastian Vettel famously named all of his cars. Some of the more memorable names were "Kate's Dirty Sister," "Hungry Heidi," and "Luscious Liz." He said it gave his cars personalities—almost like teammates with their own quirks.

Fernando Alonso reportedly clings to a very old superstition: lucky underwear. Yes, he's admitted to wearing the same pair on race days, washed of course, but still the same. Fans often joke that when Alonso finally retires, those underpants should go in a museum.

Funny Radio Messages

Team radios are supposed to be serious tools, but they've given fans some comedy gold. Kimi Räikkönen is the king of blunt replies. Once, when his engineer kept reminding him about settings, he snapped: "Leave me alone, I know what I'm doing!" It became an instant internet meme and even ended up on T-shirts.

In another race, Räikkönen told his team, "Yes, yes, yes, I'm doing all the tires," sounding more like he was ticking chores off a list than racing at breakneck speeds.

Daniel Ricciardo has the opposite vibe. He's known for laughing during races, cracking jokes mid-corner, and once yelled, "This is the best day of my life!" while charging to a podium. It's like he forgets there are millions of people listening in.

One time, when Romain Grosjean's car was on fire, he calmly radioed, "I think the car is on fire... yep, definitely burning," in the most casual tone imaginable. Fans couldn't believe how relaxed he sounded.

Food and Drink Oddities

What do F1 drivers eat? You'd expect something super high-tech, but sometimes it's surprisingly simple. Jenson Button swore by porridge with honey before every race, while Valtteri Bottas has been spotted drinking strong black coffee minutes before jumping into the car.

Drivers often burn 1,500–2,000 calories during a single race, so their post-race meals are epic. Pizza, burgers, and massive plates of pasta are common, because they need to refuel after sweating out so much energy.

Kimi Räikkönen became legendary for skipping a royal meeting with the King of Spain in 2006 because he wanted to go partying on his yacht. Instead of bowing to royalty, he was bowing to the dance floor. Fans weren't even shocked—"That's just Kimi."

Meanwhile, Daniel Ricciardo invented the grossest celebration in racing: the "shoey." After a win, he pours champagne into his sweaty racing shoe and drinks it. Even grosser, he sometimes makes other drivers or celebrities join him. It's disgusting, but it's become one of F1's funniest traditions.

Weird Car Habits

Some drivers treat their cars like living creatures. Ayrton Senna often patted his car after a victory, like thanking a horse for a good ride.

Others talk to their cars during races. Sebastian Vettel has been caught on radio whispering encouragements, as if the machine needed a pep talk.

And then there's the superstition of tapping certain buttons or switches in a specific order before take-off. Engineers roll their eyes, but drivers swear these little rituals give them focus.

Unlikely Mishaps

Formula 1 isn't always serious and smooth. In 2008, Lewis Hamilton drove into the back of Kimi Räikkönen's car at a red light at the end of the pit lane. Yes—a red light, just like at a traffic stop. The collision knocked them both out of the race, and fans joked he needed to retake his driving test.

In 2006, Juan Pablo Montoya crashed out of a race while fiddling with his drink tube. Imagine trying to sip water at 200 mph, then realizing you've lost control of your car. Oops.

And sometimes, it's just clumsy exits. In 1991, Nigel Mansell stalled his car while waving to fans after the finish. Marshals had to help push the world champion's car back into the pits, while fans laughed at the sight.

Clothing Calamities

Fireproof race suits are incredible pieces of technology, but they can also be uncomfortable. Drivers often complain that by the end of a hot race, their suits feel like soggy towels. Mechanics sometimes joke that they could wring out a liter of sweat after every race.

In 2005, Juan Pablo Montoya ripped his race suit when climbing out of his car, exposing his bright-colored underwear on live TV. Fans teased him for weeks, and he laughed it off by saying at least his underwear was fireproof too.

Sometimes the zippers on the suits break at the worst moments. One driver had to walk onto the grid holding his suit closed with tape until the team found a replacement. It wasn't the fastest-looking outfit, but at least he was legal to race.

Bizarre Celebrations

F1 podiums aren't just for trophies—they're stages for weird traditions. The champagne spray started in the 1960s when a driver accidentally shook his bottle too hard. It caught on instantly, and now no podium is complete without drivers drenching each other and the crowd.

In 2016, Daniel Ricciardo took it to the next level with his "shoey"—drinking champagne out of his sweaty shoe. He even convinced celebrities like Gerard Butler and Lewis Hamilton to join him. Fans cheered, but you can bet it tasted horrible.

Kimi Räikkönen once left the podium before the national anthem had finished playing, saying he had to use the bathroom. Broadcasters caught the whole thing, and it became one of his funniest "Iceman" moments.

Pit Lane Laughs

Pit stops are usually lightning fast, but mistakes can be hilarious. In 2011, Jenson Button's crew accidentally put the wrong tires on his car—two of one type and two of another. The mix-up ruined his race but gave fans a rare chance to laugh at the usually flawless McLaren crew.

In 2008, Felipe Massa dragged an entire fuel hose down the pit lane after being released too early. Sparks flew as it clattered behind him, and marshals chased him down to rip it free. Embarrassing, yes, but unforgettable.

And in one of the oddest moments, a team once forgot to bring the tires out at all. The driver arrived, stopped perfectly, and sat waiting while his confused mechanics sprinted back into the garage to grab the missing wheels.

Odd Car Problems

Cars sometimes fail in the strangest ways. In 2010, Sebastian Vettel's steering wheel display glitched and showed cartoon symbols instead of data. He joked afterward that it looked like a video game.

In 2006, a car's seatbelt came loose mid-race, forcing the driver to pit for a quick re-buckle. He had to sit patiently while his crew leaned over him like parents strapping a toddler into a car seat.

Occasionally, drinks systems malfunction too. Drivers press a button to sip water through a tube, but sometimes it delivers boiling-hot liquid because the car's heat warms the bottle. One driver spat it out mid-race, saying it was like drinking soup inside a sauna.

Animals on Track

Animals seem to love F1 circuits as much as drivers. In 2011, a dog ran across the track in practice in Turkey, dodging cars that were flying past at 250 km/h. Marshals eventually caught it, but not before it gave drivers a huge scare.

Groundhogs are regular visitors at the Canadian Grand Prix in Montreal. Robert Kubica once had to dodge one in qualifying, and other drivers have joked that the animals treat the track like their personal playground.

Birds aren't safe either. At the 2016 Austrian Grand Prix, a seagull nearly caused a crash when it landed on the racing line. Sebastian Vettel swerved to avoid it, later joking that he'd saved the bird's life at full speed.

Quirky Driver Habits

Ayrton Senna was known for deep focus, but also for quirks. He often wore his gloves hours before a race and would sit silently in the car long before it was time to start, almost meditating with the engine off.

Fernando Alonso has admitted to doing stretches in unusual places. Mechanics once caught him warming up in the team bathroom, balancing on tiptoes against the sink before qualifying.

Niki Lauda used to chew gum constantly during races, saying it helped him concentrate. Fans joked he must have set records not just for wins but for the world's strongest jaw muscles.

Awkward Moments

Sometimes drivers get it wrong in front of the whole world. In 2006, Kimi Räikkönen skipped a ceremony with football legend Pelé because he "didn't hear" it was happening. Cameras found him sipping a soda instead of standing on stage, and fans laughed at his classic Iceman nonchalance.

Lewis Hamilton once accidentally parked in the wrong team's pit box during practice. Rival mechanics gave him confused stares before waving him on. He later laughed, saying, "At least they were friendly about it."

In 2017, Sebastian Vettel crashed into a barrier during a demonstration run in front of thousands of fans in London. The slow-speed slip-up was harmless but unforgettable.

Unusual Crashes

Not all accidents look spectacular. In 2010, Kamui Kobayashi crashed into a barrier after his steering wheel broke in his hands.

Luckily, he was fine, but the image of a driver holding half a wheel while the other half stayed attached was bizarre.

In 2019, George Russell's car was wrecked—not by him, but by a drain cover that came loose on track. It sucked up into the underside of his car, smashing it to pieces. Marshals had to inspect every drain on the circuit before racing continued.

And in 2005, Michael Schumacher parked his car at Monaco in a suspicious way that blocked rivals during qualifying. Officials ruled it was intentional and stripped his lap time. The "crash" was less about danger and more about drama.

Gross but Funny

When drivers get dehydrated, their hands can cramp so badly that they struggle to wave after the race. One driver joked that he could barely lift his arm on the podium without looking like a robot.

Drivers' balaclavas, worn under helmets, often get drenched in sweat and reused throughout the weekend. Mechanics sometimes joke that the balaclavas could "stand up on their own" by Sunday.

And of course, there's Ricciardo's shoey—the grossest but funniest celebration in F1. He once made Sir Patrick Stewart, the legendary actor, drink from his shoe on the podium. Stewart didn't even flinch, saying, "That was delicious!" The crowd roared with laughter.

Activity – Funniest Fact Face-Off

You've just read dozens of weird, gross, and funny F1 facts. Here's a game to play with a friend or family member:

1. Each of you picks the **funniest fact** from this chapter.

2. Say your choice out loud and explain why it's the funniest.

3. The other person can challenge your choice with their own argument.

4. If you can't agree, flip a coin—but both of you have to act out the fact you picked.

Example: Pretend you're Ricciardo doing a shoey (with water, not actual shoe juice!) or Räikkönen saying "Leave me alone, I know what I'm doing."

Whoever gets the most laughs wins the "Funniest Fact Champion" title!

Chapter Seven

Rivalries and Drama: When Things Get Heated

Senna vs. Prost – The Professor and the Genius

One of the most famous rivalries in Formula 1 history was Ayrton
Senna against Alain Prost. Senna was fiery and emotional; Prost was

calculating and calm. When they were teammates at McLaren in the late 1980s, sparks flew both on and off the track.

In 1989 at Suzuka, Prost and Senna collided at a chicane. Prost walked away, thinking the race was his, but Senna restarted and still won—only to be disqualified by officials. The controversy lit a fire under their feud.

The very next year, at the same Suzuka circuit, Senna deliberately collided with Prost at the first corner to secure the championship. Fans were divided: some called it ruthless, others called it brilliant. Either way, it cemented their rivalry as the most explosive in F1.

Schumacher vs. Hill & Villeneuve – The Master of Mind Games

Michael Schumacher didn't just win titles—he often rattled his rivals' nerves. In 1994, his title fight with Damon Hill ended with contact in Adelaide, leaving Hill out and Schumacher crowned champion. Some called it opportunistic; others said it was pure aggression.

In 1997, Schumacher tried a similar move against Jacques Villeneuve at Jerez. This time, it backfired—his car ended up in the gravel while Villeneuve went on to take the title. The incident earned Schumacher a disqualification from the championship standings, proving that mind games sometimes go too far.

Schumacher's intensity made him both admired and feared. Rivals often said racing him felt like psychological warfare as much as wheel-to-wheel combat.

Teammate Wars

Sometimes the biggest drama doesn't come from opponents on other teams but from teammates. At Mercedes in 2016, Lewis Hamilton and Nico Rosberg's friendship dissolved as they fought for the title. Their collision in Spain that year, both cars out on lap one, symbolized just how far their rivalry had gone.

Sebastian Vettel and Mark Webber had a frosty relationship at Red Bull. In Malaysia 2013, Vettel ignored team orders—codenamed "Multi-21"—and overtook Webber, sparking a frosty handshake and the now-famous quote: "Multi-21, Seb, yeah, Multi-21."

Even Ferrari legends fought within their garage. Gilles Villeneuve and Didier Pironi clashed in 1982 after Pironi disobeyed team orders, stealing a win from Villeneuve. The betrayal was so deep that Villeneuve refused to speak to him again.

Heated Words and Cool Comebacks

Formula 1 isn't just about crashes and checkered flags—it's about sharp tongues too. In 2007, after a tense qualifying session, Fernando Alonso blocked Lewis Hamilton in the pit lane, sparking a feud inside McLaren. Alonso later described the atmosphere as "toxic," and the team crumbled under the pressure.

Kimi Räikkönen, usually ice-cool, gave fans one of the funniest blunt answers in 2006. When he missed a ceremony with football legend

Pelé, reporters asked why. His reply? "I was in the bathroom." Fans laughed, saying it was the most honest excuse in racing history.

Nelson Piquet, another outspoken driver, once poked fun at rivals in interviews and press conferences, sometimes stirring up more drama with words than with overtakes on track. He proved that rivalries could extend beyond lap times and into microphones.

Hamilton vs. Rosberg – Friends Turned Foes

Lewis Hamilton and Nico Rosberg grew up as friends and even shared hotel rooms when they were teenagers racing karts. But when they became Mercedes teammates fighting for championships, the friendship didn't last.

In 2014 at Bahrain, the two battled side by side for lap after lap, wheels nearly touching at 300 km/h. Hamilton won, but Rosberg's determination was clear.

By 2016, the rivalry had turned so tense that in Spain, they collided on the very first lap, knocking both cars out. Rosberg went on to win the championship that year and then retired immediately, saying he had achieved his dream and didn't want to relive the stress.

Vettel vs. Webber – The Multi-21 Showdown

Red Bull teammates Sebastian Vettel and Mark Webber never fully got along. Webber thought Vettel got preferential treatment, while Vettel believed he earned his chances by being faster.

The rivalry reached its peak in Malaysia in 2013. Webber was leading when the team told both drivers to hold position. The code "Multi-21" meant car number two (Vettel) should stay behind car number one (Webber). But Vettel ignored the message and overtook anyway.

The awkward podium afterward was legendary—Webber icy and silent, Vettel half-apologetic, and the team stuck in the middle of a public feud. Fans still quote "Multi-21" today whenever teammates clash.

Hamilton vs. Alonso – McLaren Meltdown

In 2007, McLaren paired rookie Lewis Hamilton with reigning champion Fernando Alonso. Many expected Alonso to dominate, but Hamilton shocked the world by matching him blow for blow.

The rivalry boiled over during qualifying in Hungary. Alonso deliberately delayed his exit from the pit box, preventing Hamilton from completing a final lap. The team atmosphere exploded, and both drivers lost trust in each other.

The feud got so bad that McLaren lost harmony altogether. In the end, Kimi Räikkönen snatched the championship that year, partly because McLaren's two stars had spent so much energy fighting each other instead of the competition.

Ferrari Fireworks – Sibling Rivalries in Red

Ferrari, the most famous team in F1, has seen plenty of drama between its own drivers.

In 2019, Charles Leclerc and Sebastian Vettel collided in Brazil while fighting each other for position. Both cars were destroyed, and Ferrari had to explain to millions of fans why their drivers had taken each other out.

Earlier, in the 1980s, Gilles Villeneuve and Didier Pironi had their infamous feud. After Pironi broke a "gentleman's agreement" to hold position, Villeneuve felt betrayed. It was a clash of honor versus ambition that left a permanent scar in Ferrari history.

And in 2020, Vettel and Leclerc once again collided, this time in Austria on the first lap. Fans shook their heads—it wasn't the first time Ferrari drivers forgot they were supposed to be teammates, not enemies.

Other Rivalries Worth Remembering

Nigel Mansell and Nelson Piquet had a fierce rivalry at Williams in the late 1980s. Piquet, the crafty Brazilian, and Mansell, the aggressive Brit, pushed each other to the limit. Their battles kept fans glued to every race.

James Hunt and Niki Lauda's 1976 rivalry was so legendary it inspired the Hollywood movie *Rush*. Hunt was wild and glamorous, Lauda serious and precise. Their contrasting styles made the season unforgettable.

Even within smaller teams, rivalries emerge. At Force India in 2017, teammates Sergio Pérez and Esteban Ocon collided multiple times, costing the team valuable points. Their radio messages grew spicier with each crash, turning them into one of the most dramatic mid-field duos.

Feuds That Lit Up the Grid

Lewis Hamilton and Sebastian Vettel clashed not only on track but also in headlines. In 2017 at Baku, Vettel accused Hamilton of brake-checking him during a safety car period and deliberately bumped into the back of Hamilton's car. Tempers flared, but both eventually laughed it off, realizing the sport needed rivalry as much as speed.

Max Verstappen and Charles Leclerc brought a new generation of drama. In Austria 2019, Verstappen muscled past Leclerc with a daring move that pushed him off the track. Leclerc fumed, fans debated, and the stewards eventually ruled it hard but fair racing. Their rivalry continues to keep modern F1 exciting.

Verstappen and Lewis Hamilton's 2021 title fight was one of the most intense in decades. Their season-long duel included wheel-to-wheel clashes at Silverstone and Monza, with both drivers accusing the other of recklessness. The season ended in Abu Dhabi with a hugely controversial finish, still debated by fans today.

Words Louder Than Engines

Sometimes the drama isn't in the driving but in the interviews. Nelson Piquet once caused outrage by making jokes about Nigel Mansell's family, proving that psychological battles could sting as much as on-track crashes.

Fernando Alonso famously called his Renault car a "GP2 engine" over the radio when frustrated with Honda's lack of speed in 2015. The blunt insult embarrassed Honda on their home track in Japan and went viral among fans.

Even calm drivers like Jenson Button have had their moments. After being crashed into by another driver, he once radioed: "I've just been hit by a complete rookie!" His polite but sarcastic tone made it even funnier.

Rivalries With a Smile

Not every rivalry is bitter. Daniel Ricciardo and Max Verstappen had several clashes as Red Bull teammates, including a double crash in Azerbaijan 2018 that took them both out. Instead of holding a grudge, Ricciardo joked later that it was like "two bulls fighting in a china shop."

Valtteri Bottas once gave Hamilton the nickname "Lucky Lewis" after Hamilton escaped penalties in a few tight calls. It was a cheeky jab, but Bottas and Hamilton remained friendly compared to some of the sport's fiercer rivalries.

Even rivals often become friends again years later. Prost and Senna, once bitter enemies, reconciled before Senna's death in 1994. Senna's public thanks to Prost during his final race was one of F1's most emotional moments.

Activity – Rivalry Re-Write

Imagine you're a Formula 1 journalist covering a big rivalry. Choose **two drivers from this chapter** (past or present) and write a short newspaper headline and story snippet about their clash.

Example:
Headline: "Hamilton and Rosberg Collide Again – Mercedes Garage in Shock"
Snippet: "The two former friends turned rivals made contact on lap one, leaving the team boss with his head in his hands."

Be creative! You can make your version funny, dramatic, or even exaggerate like a tabloid. Share it with a friend and see whose headline gets the most laughs or gasps.

Chapter Eight

The Tech Behind the Cars: Secret Gadgets & Smart Engineering

Aerodynamics: The Invisible Superpower

Formula 1 cars don't just cut through the air—they fight with it. The bodywork is shaped to push air in some directions and suck it away in others, creating a force called downforce. At high speeds, downforce presses the car onto the track so hard that the tires can grip like glue. Engineers like to say that if you drove an F1 car upside down through a tunnel at full speed, the downforce would keep it stuck to the ceiling. It's never been tried, but the math says it could work.

The front and rear wings are the most obvious tools in this aero-battle. They look simple, but each wing is an intricate sculpture, tested for months in wind tunnels. At high-speed tracks like Monza, engineers shave the wings down to razor-thin blades to reduce drag. At twisty tracks like Monaco, the wings are tilted steeply to act like giant fans pressing the car down. Choosing the wrong setup could mean zooming past rivals on straights but sliding helplessly in corners—or the other way around.

At the back of the car, the diffuser is the hidden hero. It's the tunnel-shaped area underneath the rear end, and it manipulates airflow so that the air rushes faster under the car than over it. That speed difference creates suction, pinning the car to the track. Teams often guard diffuser designs like treasure—rivals have been known to sneak photos when mechanics aren't looking, because one clever diffuser shape can be worth tenths of a second per lap.

Clever Gadgets

The DRS, or Drag Reduction System, is one of Formula 1's flashiest gadgets. With the push of a button, a flap in the rear wing flips open, reducing drag and giving the car an instant speed boost of 10–12 km/h. But drivers can't just spam it whenever they want—it's only allowed in designated zones, and only if they're within one second of the car ahead. This keeps racing exciting without making overtakes too easy.

Brake-by-wire is another invention that sounds futuristic. In normal cars, when you press the brake pedal, hydraulic fluid squeezes the brake pads. In F1, pressing the pedal sends a signal to a computer, which decides how to split braking between the mechanical brakes and the hybrid system that recovers energy. It's smoother, safer, and so efficient that sometimes the computer reacts faster than the driver's own reflexes.

The steering wheel is practically a spaceship control panel. With up to 30 buttons, switches, and rotary dials, drivers can adjust brake balance, fuel mix, tire settings, differential, engine power modes, and radio volume—all while taking corners at over 250 km/h. Each wheel costs around $50,000 to build. Drivers often spend hours in simulators just memorizing which button is where, so they don't accidentally switch the car into the wrong mode mid-race.

Materials and Secrets

Carbon fiber is the skeleton of an F1 car. It's made by weaving together thin strands of carbon, then baking them in resin until they become ultra-strong plates. It's five times stronger than steel but weighs much less. In crash tests, carbon cockpits survive forces that would crumple ordinary cars like soda cans.

Because every gram matters, teams obsess over weight savings. Engineers sometimes sand off layers of paint from the bodywork. A single layer of paint might weigh only a few hundred grams, but over the whole car it adds up. That's why some teams run their cars in bare black carbon fiber instead of bright colors—performance over style.

Tires are another misunderstood miracle. A Formula 1 tire is designed to last anywhere from 15 to 50 laps depending on its softness. Softer tires grip better but wear faster; harder tires last longer but don't stick as well. To get maximum grip, teams heat tires in electric blankets before races. The sweet spot is around 100°C, hotter than boiling water. Too cold and they slide; too hot and they blister. Drivers often say managing tire temperature is as much art as science.

Smart Energy

Modern F1 engines are hybrids—half traditional, half electric. The Energy Recovery System (ERS) captures energy in two ways: from braking (like regenerative brakes in electric cars) and from the turbocharger, which normally wastes energy as heat. This stored power gives drivers a "push-to-pass" boost of about 160 extra horsepower when they need it most.

Road cars now use versions of this tech. The Ferrari LaFerrari, Porsche 918 Spyder, and McLaren P1 all have hybrid systems directly inspired by Formula 1. What starts as an experiment on the track often becomes a feature in the cars we see on highways years later.

Brakes Hotter Than Fire

F1 brakes are not ordinary discs—they're made of carbon-carbon composites that can withstand temperatures above 1,000°C. That's hotter than molten lava. You can actually see them glowing red at night races like Singapore.

Because the brakes get so hot, the wheels need special ducts to funnel air onto them for cooling. If those ducts get blocked by debris, the brakes can fail within laps. Drivers have described the smell of over-heated brakes as "burning metal mixed with fireworks."

Each braking zone is a violent experience. Drivers go from 330 km/h to 80 km/h in just two seconds, experiencing forces up to 5 g. That's like having five times your body weight crush your chest every time you slam the pedal.

Fuel for Rockets on Wheels

The fuel in an F1 car is similar to normal gasoline but with a secret recipe. It's blended for maximum efficiency and power while still meeting strict regulations. Engineers describe it as "the cleanest fuel in the world."

Refueling was once part of F1 strategy, but it was banned in 2010 because of safety concerns. Fireballs in the pit lane were far too common. Now cars start with a full tank and must manage fuel carefully across the whole race.

Fuel loads also shape race strategies. A heavy tank makes the car slower but means fewer pit stops; a lighter tank makes the car faster but requires earlier pit stops for tires. Finding the sweet spot is like solving a complex math puzzle while racing at 300 km/h.

Suspension and Ride Tricks

The suspension in an F1 car doesn't look like normal car suspension. Instead of springs and shocks, it uses a complex system of wishbones, torsion bars, and dampers, all packed tightly into the chassis.

Active suspension—where computers controlled the car's ride height—was briefly allowed in the early 1990s. Williams used it to dominate, but it was banned for making cars almost "too perfect." Still, elements of the tech live on in how modern suspensions are tuned.

Tiny ride height adjustments can transform a car. Even a few millimeters higher or lower can mean either skimming along smoothly or scraping dangerously against curbs. Teams use lasers to measure this down to the millimeter.

Data Rules Everything

Every F1 car is packed with sensors—over 300 of them—measuring everything from tire pressure to oil temperature to the g-forces on the driver. During a race, the car streams gigabytes of data back to the garage every second.

Engineers sit in the pit wall and in giant team headquarters thousands of kilometers away, analyzing this river of data live. They can spot problems before the driver even notices them. Sometimes a voice comes over the radio: "Box, box, box," just as a part is about to fail.

Telemetry also reveals how drivers behave. Teams can see exactly how late each driver brakes, how hard they turn, and how early they accelerate. It's like having a video game replay of every lap, down to the tiniest detail.

Hidden Tricks and Secrets

Formula 1 engineers are famous for bending the rules. In 2009, Brawn GP invented the "double diffuser," a clever design that created huge downforce from the floor of the car. Rivals called it illegal, but officials ruled it was within the rules. Brawn went on to win both championships that year.

Mercedes introduced the DAS system in 2020—Dual Axis Steering. By pulling or pushing on the steering wheel, drivers could adjust the angle of their front tires on straights, improving tire warm-up and reducing drag. It was ingenious, but the FIA quickly banned it after that season.

Sometimes tricks are simpler. Teams have even experimented with "flexi-wings"—wings that bend just enough at high speeds to reduce drag, while appearing rigid in inspections. Spotting and banning these sneaky inventions is a constant game of cat and mouse between engineers and rule-makers.

Safety Tech That Saves Lives

The "survival cell" is the strongest part of the car. It's a carbon-fiber tub around the driver, tested by firing it into walls at terrifying speeds. In crashes where the rest of the car shatters, the survival cell stays intact like a protective shell.

The halo, a titanium frame added in 2018, can support the weight of a double-decker bus. It's saved lives multiple times already, including Lewis Hamilton at Monza 2021 and Zhou Guanyu at Silverstone 2022. Some fans disliked how it looked, but today it's seen as one of the greatest safety inventions in racing.

The HANS device (Head and Neck Support) is another quiet hero. It's a carbon collar strapped to the helmet and shoulders that keeps the driver's neck safe during high g-force crashes. Without it, whiplash injuries would be far more common.

Driver Gear – High-Tech Armor

Race suits look like colorful pajamas, but they're made of Nomex, a fireproof material that can withstand flames for over 10 seconds without burning the skin. Suits are tested in ovens before they're approved for racing.

Helmets are marvels of engineering. They're made of carbon fiber and kevlar layers, weigh just 1.5 kg, and can stop bullets in testing. Each one is hand-painted with personal designs—like Daniel Ricciardo's playful artwork or Sebastian Vettel's changing designs almost every race.

Gloves even have sensors built in. Since 2018, F1 gloves can monitor a driver's pulse and blood oxygen levels during a crash, giving medical teams instant data. It's like wearing a fitness tracker that could save your life.

Futuristic Inventions

Hybrid power units are already space-age, but engineers are now working on fully sustainable fuels. The goal is for F1 to be carbon-neutral by 2030, proving that even the fastest cars in the world can be environmentally friendly.

Active aerodynamics may return in the future. Imagine wings that automatically adjust every corner to give perfect balance between speed and grip. It's banned today, but engineers are always dreaming of the next breakthrough.

Even pit stop equipment is evolving. Some teams are testing automated wheel guns with sensors that confirm a tire is secure in less than a millisecond. Soon, mistakes like loose wheels might be eliminated completely.

And then there's simulation technology. Drivers spend weeks in giant simulator pods that look like arcade games but cost millions of dollars. These simulators replicate every bump of every track and are so realistic that drivers sometimes forget they're not in a real car.

Activity – Design Your Own Gadget

Imagine you're the chief engineer of your own Formula 1 team. You've been given permission to invent **one brand-new gadget** for your car. It could be:

- A new type of steering wheel with special buttons

- A futuristic wing that changes shape mid-race

- A driver suit with cooling fans built in

- Or something completely wild, like tires that glow when they get too hot

Draw or describe your invention, give it a name, and explain how it would help your driver go faster or stay safer. Then share it with a friend or family member and see if they'd approve it for the rules—or ban it for being "too clever"!

Chapter Nine

Records, Stats, and Amazing Achievements

Champions of Champions

Michael Schumacher and Lewis Hamilton share the record for the most world championships, with seven each. That's like winning your school's sports day every single year for almost a decade. Schumacher won his five titles in a row with Ferrari from 2000 to 2004, turning the red cars into an unstoppable machine. Hamilton's run came later, between 2008 and 2020, with a mix of McLaren and Mercedes dominance. Whenever one name is mentioned, the other follows—their numbers are so alike that fans still argue who is the ultimate greatest.

Hamilton also holds the record for most race wins, with more than 100. Imagine winning a race every time you played a video game and then doubling that. To put it in perspective, many great drivers in history barely cracked double digits. Hamilton has more wins on his own than entire teams that have raced in F1 for decades.

Before these two modern titans, there was Juan Manuel Fangio. Racing in the 1950s, he won five championships with four different teams. He didn't just stick with one car—he hopped between Alfa Romeo, Maserati, Mercedes, and Ferrari, and still came out on top. Fangio was nicknamed "El Maestro" for his smooth style. His record stood untouched for nearly half a century until Schumacher surpassed him.

Masters of Speed

F1 is about speed, and the records prove it. The fastest ever speed recorded in an official F1 race is 372.6 km/h (231.4 mph), set by Valtteri Bottas at the 2016 Mexican Grand Prix. To imagine it, picture sticking your head out of a car window on the highway... then multiply that wind blast by four. Bottas was moving faster than most airplanes when they leave the ground.

Speed doesn't just mean straight lines—it also means lightning-fast pit stops. In 2019, Red Bull's pit crew changed all four of Max Verstappen's tires in just 1.82 seconds. That's faster than it takes most people to tie one shoelace. The crew practiced hundreds of times, each person shaving tenths of a second until the stop looked more like magic than mechanics.

Kimi Räikkönen holds the record for the fastest lap in F1 history, set at Monza in 2018. He averaged 263 km/h (163 mph) over the entire lap. Imagine running around your school's football field but doing it at highway speeds without ever slowing down—that's how relentless Monza is, and Räikkönen mastered it.

Iron Men of Racing

F1 drivers may seem like rock stars, but the real heroes are the ones who stay in the sport for decades. Rubens Barrichello once held the record for most race starts, with 322. That's 18 full years of lining up on the grid, traveling the globe, and strapping into some of the most powerful machines ever built. For younger fans, that's like starting kindergarten and still racing when you finish college.

Fernando Alonso has since gone even further, surpassing 360 race starts and still racing strong. He began his career in 2001, stepped away for two years, then came back to compete with drivers half his age. Fans joke that Alonso will never retire, and that one day he'll still be overtaking cars with a walking stick taped to the steering wheel.

On the other end of the spectrum, Max Verstappen made history by becoming the youngest driver to start a Grand Prix, at just 17 years and 166 days old. While most kids at 17 are nervously trying their driving test, Verstappen was racing wheel-to-wheel against world champions. He scored points in only his second race, proving he wasn't just a teenager with a license—he was a superstar in the making.

Age doesn't always mean weakness, either. Luigi Fagioli became the oldest driver to win a race in 1951, at 53 years old. At an age when most athletes are long retired, Fagioli was still outsmarting younger rivals. His win proved that in racing, brain and bravery can sometimes outdo raw youthful reflexes.

Streaks That Stunned the World

Sebastian Vettel holds the record for the most consecutive race wins in a single season: nine in a row during 2013. He didn't just win—he dominated, leading from the front so often that fans joked Red Bull's strategy meetings were basically nap time. Nine straight wins is like never losing a single level in a video game for weeks on end.

Lewis Hamilton has the record for the most consecutive seasons with at least one victory: 15 years in a row, from 2007 to 2021. That means every single year he lined up, he found a way to win at least once, no matter how strong or weak his car was. Talk about consistency.

McLaren holds the team record for the longest unbeaten run. From 1988 to 1989, they won 11 races in a row, powered by Ayrton Senna and Alain Prost. It was so one-sided that rival teams called it "The McLaren Show."

Pole Position Kings

Lewis Hamilton has started more races from pole position than any other driver—over 100 times. That means he was the fastest in qualifying and got to line up first on the grid. Starting from pole is like sitting in the front row at a concert—you get the best view and the cleanest getaway.

Ayrton Senna, however, is remembered as the "Qualifying King." He scored 65 pole positions in an era when cars weren't nearly as finely tuned as today. Senna's secret was pushing beyond the limit for one lap, wringing every ounce of speed from his car. Rivals said watching his qualifying laps was like witnessing poetry at 300 km/h.

On the team side, Ferrari has claimed the most pole positions in history. For decades, their red cars have been regular front-row starters, proving that even when they don't win championships, they're still the kings of Saturdays.

Longest and Shortest Stories

The longest Formula 1 race ever was the 2011 Canadian Grand Prix. Rain delays and safety cars stretched it to over four hours. Jenson Button started near the back, had multiple pit stops, and even collided with his teammate—but somehow came through to win on the very last lap. It was an endurance test for fans and drivers alike.

On the opposite end, the shortest race in history happened at the Belgian Grand Prix in 2021. Torrential rain made it too dangerous to race. Drivers did just two laps behind the safety car, then the race was called off. Fans who waited in the cold and wet all day were furious, but the rules meant half-points were awarded anyway.

Monaco has its own unique record: the slowest winning average speed. Because the streets are so narrow and twisty, winning there means averaging only about 150 km/h—still fast, but snail-like compared to Monza's 260 km/h averages.

Youngest and Oldest Feats

Max Verstappen, as well as being the youngest starter, is also the youngest race winner. He claimed victory in Spain 2016 at just 18 years old. Most kids his age were stressing about exams—Verstappen was spraying champagne on the podium.

On the other side, Kimi Räikkönen holds the record for the longest gap between his first and last race wins: 15 years. He first won in 2003 and was still winning races in 2018. His icy personality and longevity earned him the nickname "The Iceman."

Pit Stop Perfection

Red Bull holds the record for the fastest pit stop ever: 1.82 seconds at the 2019 Brazilian Grand Prix. That's less time than it takes to open a soda can. The crew rehearsed their movements thousands of times, like dancers perfecting choreography.

Williams, though, once set a record for the fastest *average* pit stops over an entire season. In 2016, they consistently delivered sub-2.5 second stops, proving that speed isn't about luck but about repeatable precision.

Ferrari holds the opposite kind of record: in 1999, they accidentally fitted three tires of one type and one of another on Eddie Irvine's car. It was so disastrous that it became one of the most replayed pit stop blunders ever.

Tracks and Travel

Formula 1 is truly global, and the records reflect it. The country with the most Grands Prix held is Italy, which has hosted races at Monza, Imola, Mugello, and even Pescara. Monza alone has been on the calendar almost every single year since 1950.

The longest track ever used was the old Nürburgring Nordschleife, at 22 kilometers per lap. Drivers called it "The Green Hell" because it was so dangerous. A single lap was longer than entire races at some modern circuits.

The shortest F1 track is Monaco, at just 3.3 kilometers. Drivers have to do nearly 80 laps to cover the same distance as fewer than 50 at most other venues.

Weird and Wonderful Stats

Pastor Maldonado holds the odd record of scoring just one victory in his career—but what a win it was. In 2012, he triumphed at the Spanish Grand Prix in a midfield Williams car, shocking everyone. His name became a symbol of unpredictability: anything can happen in Formula 1.

Andrea de Cesaris once held the record for the most races without a win: 208. He was fast, but luck and crashes always seemed to get in the way. Fans called him "Andrea de Crasharis," a nickname he later laughed about himself.

Jarno Trulli was famous for leading long trains of cars without letting anyone past. This earned him the nickname "The Trulli Train," and it became one of the quirkiest records—leading the most laps without being the fastest car.

Activity – Record Breaker Game

- Here's a challenge for you: pick one of these record categories and invent your own *wild* version.

- **Fastest Pit Stop:** How fast could your family change all four shoes on a chair pretending it's a car? Time it!

- **Longest Race:** Create an "endurance challenge" at home—how many times can you run around your living room without stopping?

- **Weirdest Record:** Make up a funny F1 stat, like "most bananas eaten before qualifying" or "fastest shoelace tie on the grid."

Write down your record and share it with a friend. See if they can beat it—or come up with a crazier one of their own!

Chapter Ten

The Future of Formula 1: What's Next?

Greener, Cleaner, Faster

Formula 1 has set itself a bold goal: to be **carbon neutral by 2030**. That means every bit of pollution created by cars, planes, and trucks will either be reduced or balanced out. It's no small task—F1 cars race all over the globe, and the sport ships tons of equipment each week. To reach this goal, teams are redesigning factories to use renewable energy, and organizers are planning smarter schedules to cut down on unnecessary freight miles. Imagine trying to move an entire circus across continents, but with an eco-friendly twist.

The fuel of the future is one of F1's most exciting experiments. By 2026, the cars will run on **100% sustainable fuel**. Instead of being dug out of the ground, this fuel will be made from renewable sources like plants, waste materials, or even captured carbon pulled straight out of the air. Engineers believe that if F1 can make eco-fuel work at 300 km/h, the same fuel could also power ordinary cars and trucks. Think of it as Formula 1 becoming a laboratory for the whole planet's future.

Hybrid engines aren't going away either. Right now, they recover energy when drivers brake and from heat in the turbochargers. But the next generation of engines will take it further—harvesting more energy, wasting less heat, and squeezing out extra horsepower without using a drop more fuel. The dream? To make engines smaller, greener, and even more powerful at the same time. Road cars will likely borrow these tricks, so your family's next hybrid might secretly have a little F1 DNA.

Smarter Cars

Formula 1 bans many "driver aids" because the sport insists on keeping the humans in control. That's why there's no cruise control, no lane assist, and no automatic braking like you see in road cars. But technology is creeping in around the edges. Some systems already help drivers optimize fuel use or energy recovery. In the future, cars may have more intelligent onboard software that fine-tunes performance in real time, acting like a digital co-pilot while still leaving the big decisions to the driver.

Aerodynamics, the art of shaping air, could also leap forward. Right now, only the DRS wing flap can move mid-race, but engineers are exploring **active aero**—bodywork and wings that change shape for each corner. Picture a car flattening its wings for the straights, then sprouting new shapes for a tight bend, like a Transformer. If it's ever legalized, it could make cars faster, but critics argue it might feel like racing robots instead of humans.

Even tires might get a radical upgrade. Pirelli, F1's tire supplier, is experimenting with **airless tires**. These don't use traditional air pressure, so they can't get punctures. That would mean no more races ruined by flat tires. But there's a catch: they need to feel the same as current tires so drivers can still sense grip levels. If they succeed, it could change racing—and everyday driving—forever.

Digital Racing

Formula 1 already relies on simulators, but they're about to get even bigger. Drivers spend weeks in giant motion pods that mimic every bump of every track. In the future, augmented reality (AR) and virtual reality (VR) could bring those simulations to fans. Imagine wearing AR glasses during a live race and seeing tire temperatures glowing red on the cars in real time, or driver heart rates flashing above their helmets like a video game HUD.

Esports is also exploding. The F1 Esports Championship now draws millions of viewers, and some virtual racers have made the leap into real racing. One day, a world champion might come not from karting but from gaming. The barrier between the digital track and the asphalt one is shrinking every season.

Artificial intelligence is creeping into strategy too. Modern races already have engineers crunching numbers to predict the best pit stop timing. AI could turbocharge that, analyzing millions of possible scenarios in seconds and whispering the perfect choice to the team. The big question: will instinct from a human strategist beat raw computer brainpower, or will AI become the secret weapon that decides championships?

Fan Experience of the Future

Formula 1 races already attract hundreds of thousands of fans in person, but organizers want to take it further. In the future, tracks could be surrounded by interactive zones where fans try simulators, race drones, or even test electric go-karts between sessions. The idea is to make race weekends a festival, not just a race.

At home, fans may soon control their camera angles during live broadcasts. Imagine watching an onboard view from your favorite driver for the entire race, while your friend watches from the perspective of pit lane. Interactive streaming is being tested, and F1 wants to give fans the choice of their "personal race."

Holograms might also play a part. Some companies are experimenting with projecting 3D race cars into living rooms or fan zones. One day, kids might watch a car zoom around their coffee table while the real version blasts around the track.

Expanding Around the World

Formula 1 already races on five continents, but the future could see even more global expansion. New events in Las Vegas, Miami, and Saudi Arabia have shown how fast the calendar can grow. Rumors swirl about potential races in South Africa, returning to Kyalami, and maybe even a future Grand Prix in countries like India or Vietnam.

Night races are another trend. Singapore proved how stunning they look, and more circuits may follow, with glowing city skylines as backdrops. Floodlit tracks not only look dramatic but also allow viewers in different time zones to watch at convenient hours.

"Street circuits" are becoming more common too. Instead of traditional racetracks, F1 cars zoom through city centers. It makes the sport more accessible for fans, who don't need to travel to remote tracks. Expect more cities to bid for a spot on the calendar in the years ahead.

Safer Than Ever

Even though F1 is already incredibly safe, the future promises even more protection. One area is **helmet tech**. Future helmets may have heads-up displays (HUDs), showing drivers lap times, warning signals, or even a mini-map directly inside the visor. The challenge is adding this without distracting the driver.

Fireproof clothing will also keep improving. Scientists are developing lightweight, heat-resistant fabrics that are thinner but even stronger. That means drivers will stay cooler in tropical races but just as safe in emergencies.

Medical response may also get faster. Drones could one day carry first-aid gear to crash sites in seconds, beating medical cars through crowded circuits. Combined with biometric gloves that already measure a driver's health in real time, medics will know exactly what's wrong before they even arrive.

Data and Connectivity

Right now, F1 cars beam data to their garages instantly, but the future might allow fans to access more of it live. Imagine pulling up a phone app that shows every driver's fuel level, tire condition, and engine health in real time.

5G networks will make data transfer almost instant. This could allow engineers at team headquarters—thousands of kilometers away—to monitor cars as if they were standing trackside. Teams are already experimenting with "remote garages" where half the engineers never leave their home country.

For drivers, future radio systems might include translation features. If a driver speaks one language and an engineer another, AI could translate instantly so both understand perfectly, reducing confusion in the heat of battle.

Futuristic Cars

Electric cars dominate road car headlines, but Formula 1 insists it will keep roaring engines—just with cleaner fuel. Still, many wonder if a fully electric F1 will arrive someday. For now, Formula E explores that territory, but F1 may borrow features like faster charging and quieter technology.

Self-healing materials could be the next big leap. Imagine a front wing that repairs its own cracks mid-race, like skin healing after a scratch. Scientists are already testing polymers that do this. If F1 adopts them, broken wings might soon be a thing of the past.

Lightweight batteries are also on the horizon. Current hybrid systems are heavy, but engineers are developing smaller, more powerful cells. A lighter battery means more speed without sacrificing energy.

Fans in the Driver's Seat

Virtual reality may one day allow fans to ride inside the car—without leaving their couch. Picture wearing a VR headset and feeling like you're sitting in Lewis Hamilton's cockpit at 300 km/h, hearing his radio messages and seeing what he sees.

F1 could even experiment with fan voting. Imagine an online poll during a race where fans choose the "Driver of the Day" (already in place) or maybe even vote on fun non-crucial things like helmet designs or trophy shapes. It's a way to make spectators part of the show.

Esports might go even further, with hybrid events where gamers race virtually alongside real F1 drivers. The lines between "pro" and "player" are already blurring. One day, the kid dominating on a home console could find themselves drafted by a real team.

The Global Show

F1 might expand into more countries than ever before. Africa is the only continent (besides Antarctica) without a current race, but South Africa is a strong candidate for a comeback. Imagine the sound of engines roaring through Kyalami while zebras graze nearby—it would be unforgettable.

Sustainability goals also mean race weekends might look greener. Solar-powered paddocks, reusable grandstands, and even plant-based catering are already being discussed. The future race might be as much about eco-friendly innovation as it is about who crosses the line first.

There's even talk of hosting races on artificial islands or floating circuits. Engineers in the Middle East have suggested water-based grandstands, where fans watch from yachts and floating platforms. It sounds like science fiction, but then again, so did night races before Singapore made them real.

Activity – Future F1 Inventor

You've seen how F1 is changing—smarter cars, greener fuel, and futuristic fan ideas. Now it's your turn to imagine the future.

> 1. Invent **one new feature** for F1.

- Maybe a car that changes colors when it heats up.

- Or helmets with glowing visors.

- Or races on the Moon with no gravity!

2. Draw or write a short description of your idea.

Share it with a friend or family member, and ask: would this make F1 more fun, or completely crazy?

Your invention might be silly, brilliant, or both. After all, today's "impossible" idea often becomes tomorrow's record-breaking technology.

Printed in Dunstable, United Kingdom

73685120R00063